Salim Ali

INDIA'S BIRDMAN

Published in 2003 by

Rupa & Co

7/16, Ansari Road, Daryaganj
New Delhi - 110 002

Sales Centres:
Allahabad Bangalore Chandigarh Chenni
Dehradun Hyderabad Jaipur Kathmandu
Kolkata Ludhiana Mumbai Pune

Photographs, courtesy: Bombay Natural History Society

Dr. Salim Ali Papers, courtesy: Nehru Memorial Library, New Delhi

Cover & Book Design by
Shashi Shetye
307 Sector 8, Pocket C-8, Rohini,
Delhi 110085.

Printed in India by
Gopsons Papers Ltd.,
A-14 Sector 60
Noida 201 301

Salim Ali

India's Birdman

Reeta Dutta Gupta

Rupa & Co

CONTENTS

For my precious parents
Nripen and Bulbul

With a Siberian Crane chick. International Crane Foundation, U.S.A.
(Photo: George Archibald)

Chapter One

Life and Times

A story goes that a journalist was interviewing Dr. Salim Ali, the doyen of Indian ornithology, when the Grand Old Man, who was hard of hearing, interrupted frequently saying 'pardon'. But suddenly, his face lit up and pointing to a tree in the distance, he exclaimed, "Did you hear that? That was a barbet!"

Such was Salim Moizuddin Abdul Ali's involvement with birds that his ears and his eyes were always open to the call of birds and their songs and to their presence in thickets, bushes, hedges and shrubbery and on tree tops, even when he was not engaged in ornithological activities. And when he was, he remained totally engrossed in his winged beauties. He spent over seventy long years studying birds and every time he saw one, even if it was the ubiquitous house crow or the pugnacious bulbul, he looked at it with such rapt attention as if he was seeing it for the first time. "You forgot about most things when you are concentrating on birds. You enjoy it. It's like looking at a beautiful woman!" he had stated in an interview.

Once this concentration, which made him 'unmindful' of other things, nearly cost him his life in Tibet. He wrote in his autobiography *The Fall of a Sparrow*:

"I recall one particularly hair-raising incident…on my way to Lake Manasarovar and Mt Kailas in 1945… It was at a particularly narrow part of the trail with a thousand feet of vertical scarp on one side and the roaring Kali river some 300 feet vertically down on the other. I had walked ahead of the porters… Just at that moment a tiny bird—how well I remember that Yellownaped Yuhina!—got up to the top of a bush, some yards away… Just as I got it in the field of my glasses, it hopped a bit further up, so to get a better view I took a step back… As I did so, I felt a small pebble slip from under my heel and heard a faint continuing clatter as it went rolling down the hill …I casually looked back… What I saw literally made my hair stand on end …two inches more and I would have followed that rollicking pebble."

From his childhood to the very end of his life, birds held a special attraction for Salim Ali and he followed his feathered friends across forests, deserts and dales and on mountaintops—to almost every nook and corner of

Did you hear the Barbets' calling?

H.R. Yadav

Yellownaped Yuhina

the country—observing them and documenting their habitats and behaviour with such single-minded dedication that he came to be called India's Birdman. And in the due course of his life, numerous awards, decorations and honours, both national and international, came his way for his distinguished service to Indian ornithology and conservation: the Asiatic Society's Joy Gobinda Law Medal for researches in Asiatic zoology (1953); Padma Bhushan (1958); three science doctoral degrees (Aligarh Muslim University, 1958; Delhi University, 1973; Andhra University, 1978); the Union Gold Medal of the British Ornithologists' Union (1967); the International Union for Conservation of Nature and Natural Resources' John C. Phillips Memorial Medal (1969); the prestigious J. Paul Getty International prize for Wildlife Conservation (1976); Padma Vibhushan (1976); International Conservation Award of the National Wildlife Federation, USA (1983); the National Gold Medal for Wildlife Conservation (1983), and many more. He was the first Indian ornithologist par excellence (even though he did not have a formal degree), who put ornithology on the map of India and made the common man aware of birds with his persuasive writings and his fine bird books. His classic work *The Book of Indian Birds*, the 'Bible' of every bird-watcher as well as the field ornithologist, has gone into its twelfth edition and has been reprinted three times.

Indira Gandhi, referring to the above work wrote in her *Foreword* to the children's book *Our Birds*: "Like most Indians, I took birds for granted until my father sent me Sri Salim Ali's delightful book from Dehra Dun jail and opened my eyes to an entirely new world. Only then did I realise how much I had been missing." Again, as the Patron of the Bombay Natural History Society (BNHS) when she was the Prime Minister, she stated in a speech: "I had always loved animals but I didn't know much about birds until the high walls of Naini prison shut us off from them, and for the first time I paid attention to bird song.

With Indira Gandhi at an award ceremony, 1976

I noted the songs, and later, on my release, was able to identify the birds from Dr. Salim Ali's book."

Commenting on the growing interest in birds among the people as reflected in the increasing number of visitors to the Keoladeo National Park near Bharatpur, the incomparable wetland that Salim Ali had helped to preserve, the ornithologist once stated, "In my younger days, I never knew any Indian interested in birds. To imagine that an Indian woman would be going about looking at birds was unbelievable. And the…Europeanised Indians used to occasionally have a look because, I suppose, they thought it was the thing to do so. But now there is real interest on a very large scale." And for this, all accolades go to the Old Man, as Salim Ali was affectionately called by his colleagues and associates.

Born on November12, 1896 and orphaned at an early age, Salim Ali was the youngest sibling of five brothers (Hasham, Hamid, Aamir, Jabir and Salim) and four sisters (Ashraf, Akhtar, Farhat and Kamoo). His mother, Zeenat-un-nissa, died when he was a child of three and his father, Moizuddin, passed away when he was just a year old. But the Ali children were fortunate, for a maternal uncle, Amiruddin Tyabji, and his childless wife, Hamida Begam, enfolded their protective arms around them and reared them with love and care in their house in Khetwadi in Bombay (now Mumbai). Uncle Amiruddin,

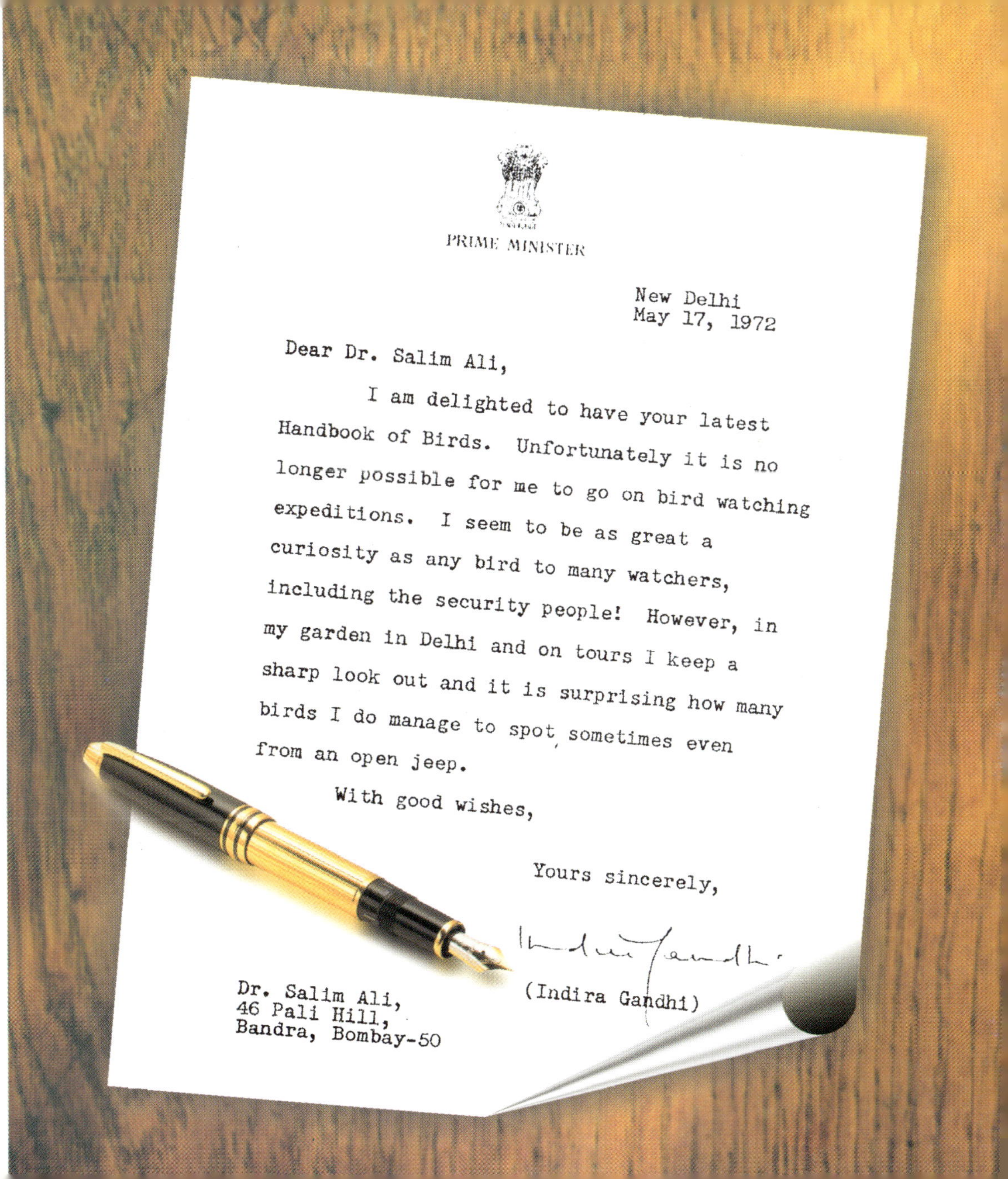

PRIME MINISTER

New Delhi
May 17, 1972

Dear Dr. Salim Ali,

I am delighted to have your latest Handbook of Birds. Unfortunately it is no longer possible for me to go on bird watching expeditions. I seem to be as great a curiosity as any bird to many watchers, including the security people! However, in my garden in Delhi and on tours I keep a sharp look out and it is surprising how many birds I do manage to spot, sometimes even from an open jeep.

With good wishes,

Yours sincerely,

Indira Gandhi

(Indira Gandhi)

Dr. Salim Ali,
46 Pali Hill,
Bandra, Bombay-50

a landed gentleman and a well-connected *shikari* (the sport of hunting carried a status symbol in yesteryears and was not yet a taboo then) initiated the boy Ali to *shikar*, which made him become aware of the natural world around him. And it was he who was instrumental in introducing the boy to ornithology.

Even as a child of nine or ten, Salim Ali showed a keen interest in birds, and according to Laeeq Futehally, his niece and co-author of the book *Common Birds*, the boy had "set up a scientific museum of squirrel skins, lizards and beetles on the dressing table." And in his memoir, Salim Ali himself tells us how he and his nephew, Suleiman, two years his junior, secretly kept some birds bought for the table as pets in an inventive cage made from wire mesh and old packing cases with the help of Nannoo, the cook. The 'aviary' gave the boys immense joy as they sat for hours together beside the enclosure watching the behaviour of birds and naming their favourite birds. Sometimes, they also bought birds from the bird market to add to their collection. "In the early days, our interest was confined chiefly to game birds, and our collection consisted of Grey and Painted Partridges, and Grey, Rain and Bush Quails. Occasionally…we would acquire a pair or two of Grey Jungefowl or Red Spurfowl. But I never was nor have been successful in keeping captive birds…"

When Salim Ali was nine, an incident—the fall of a sparrow—opened up undreamed vistas for him, making him take a serious interest in birds so that he eventually became one of the world's greatest ornithologists. Uncle Amiruddin had presented an air

gun to the boy and with it he hunted sparrows for the pot. Thankfully, out of this sport, which would be considered vandalistic and deplorable today, a positive chain of events followed.

Yellowthroated Sparrow

One day, little Ali shot a sparrow that had an uncommon patch of yellow on its throat like a 'curry stain'. "I knew one could eat the house sparrow, but I wondered whether it was lawful for a little Moslem boy to eat a sparrow with a little yellow patch, so I asked my uncle," Salim Ali recalled in an interview. Uncle Amiruddin, one of the earliest Indian members of the BNHS, (founded in 1883), sent the boy to meet the Society's secretary, Walter Samuel Millard, who at a glance not only identified the bird as a Yellowthroated Sparrow (*Petronia Xanthocollis*) but also showed him the hundreds of different birds found in the Indian Empire of the British Raj in their vast specimen collection, as well as explained to him points of differences and identification. Millard also introduced the boy to "inimitable classics" like Edward Hamilton Aitken's *Common Birds of Bombay* and *A Naturalist on the Prowl*— books to which Salim Ali remained addicted throughout his life.

In a journal of the BNHS, Salim Ali wrote, "I personally have very special cause for gratefulness to Millard for the inordinate amount of interest he took in that diffident little schoolboy who once approached him with the mangled carcass of a Yellowthroated Sparrow for identification in the year 1906 or thereabout. It was largely the fatherly

encouragement I received from him at the time, and continuing thereafter that set me off on a lifelong enjoyable and rewarding bird study." Here at the Society, he learnt the science and art of skinning, stuffing, labeling and preserving museum specimens with the help of the BNHS staff, who were in those early days mostly British naturalists.

Salim Ali went to primary school at Zanana Bible Medical Mission Girls High School (later christened Queen Mary's High School for Girls) with his sisters Akhtar and Kamoo. Later, he joined St. Xaviers', where all his brothers had studied. An average student throughout school life, the only prize he ever got at school (interestingly, a book titled *Our Animal Friends*) was in the fourth grade for 'good conduct'. School over, he enrolled himself at St Xaviers' College, but his intense dislike for a branch of mathemacitcs "logarithms and suchlike evils" made him soon dump college education and at the age of 18, he left for Burma (now Myanmar) to join J.A. Ali Bros, his brother Jabir's tin and wolfram mining business firm.

Salim Ali as a young man

Here, the backwater town of Tavoy in lower Burma was his home from 1914 to 1917 and from 1919 to 1923 with a brief break in India in-between when he got married and then took along his wife, Tehmina, to Burma. And here too he found himself looking for birds and making a private specimen collection. "The mining business was all in thick forest and forested country. That part of the country where I was, the

Tennasserim was particularly good for birds," he recalled in an interview. But he returned to India in 1923 as birds and not business was the call and destiny of his life.

A review of E.C. Stuart Baker's volume on birds in the series *The Fauna of British India* in a BNHS journal in 1925 also influenced him "only partly" as he put it, to become the first Indian ornithologist. The reviewer had lamented, "There is food, however, for reflection in the fact that we can trace no single piece of information in the two volumes which can be attributed to the work of an Indian naturalist. With the elimination of the European element in India, are we about to witness the stoppage of all scientific progress in the fertile fields of the Orient, and is this work to be the swan-song of Indian ornithology?" The Gold Medal of the British Ornithologists' Union, rarely given to a non-British, was awarded to Salim Ali in 1917, which proved the reviewer wrong.

H. Saiduzzafar

With Tehmina at Dehra Dun, 1934

In 1917, when Salim Ali returned to India for a break, he took a formal course in Commercial Law and Accountancy, but he also added to his schedule a course in zoology at St. Xaviers when Father Blatter, the Director of the Biology Department, advised him to opt for a scientific zoological background saying, "You are so keen on birds." Years later, Salim Ali recalled, "He (Father Blatter) asked me to go to college, sit with the Honours class and get myself properly trained. So that was a good opportunity and I did that for one year. I did not sit for any proper examination because I was not thinking of a degree but knowledge..." He attended the zoology class with Stanley H. Prater, who "did not have a degree either," and who later became the curator of the BNHS. "He (Prater) helped with unsavoury jobs such as dissecting a cockroach. Both of us, because we were keen ourselves managed to absorb far more than any student who did it for a degree," Salim Ali said. And soon, after the zoology class was over, they rushed back to the BNHS. There, as he noted in his autobiography, "I spent many hours several times a week in the library or rummaging among the bird collection to familiarise myself with Indian birds..."

'Chance favours the prepared mind' so wrote Louis Pasteur. Finally chance did come to Salim Ali, who was so keen on specialising on birds, when Professor Erwin Stresemann of the Berlin University Zoological Museum, whom he came to regard as his *guru*, agreed to train him. After returning from Burma in 1924, Salim Ali looked in vain for a job in the field of natural history in which his heart was set. He tried for a job with the Zoological Survey of India but the post went to another man with a formal degree. Finally, he found work for two years as a guide-lecturer at the newly opened natural history section of the Prince of Wales Museum, Bombay. But he soon tired of saying the same thing over and over again for what he wanted now was to study ornithology. Recalling his days as a guide-lecturer, he said, "Blind boys used to come to the museum and pass their hands over the various objects. They had a wonderful sense. If you gave them the skull of a cat, or the skull of a tiger, and after several other things you gave them the skull of a cat, they would put their hands over and say this is a small tiger, or just like a tiger. Then it got boring and birds was my thing and I wanted to specialise in birds. At that time there was not a single institution in India where you could do anything more than dissect a pigeon, normal zoology course."

So, in order to study birds, Salim Ali first requested the British Museum (Natural History) for training facilities and on failing to get a positive response, he wrote to Professor Erwin Stresemann, who was "absolutely the top man in the world in his field." And it was from Stresemann that he learnt systematic ornithology and he recalled, "I tried to connect the geographical features and vegetation and so on with the types of birds found there. And this is what Stresemann had put me on. He said that now that all these naming of birds and museum work is overdone, these are really the things. We know everything about the dead birds, but we know nothing about the living bird."

In another interview he stated, "He (Stresemann) was very kind indeed and took a lot of trouble and spent a lot of his time and worked with me and showed me exactly how it is all done and what the significance of various parameters were. So I got a very good grounding in about nine months with him. After that, I went to the British Museum just to see what they were doing because they had amassed a lot of Indian material...For the Indian birds, the British Museum is the best and you cannot do any work unless you consult the British Museum."

Dr. Sidney Dillon Ripley: A giant of his age and a close friend of the Birdman

If Salim Ali had to struggle through years of unemployment and hardship during his early years, he was also fortunate in having the opportunity to work with some of the noted naturalists of his time, all of whom came to hold Salim Ali in great esteem: S.H. Prater (who became a close friend and offered him the assignment of writing *The Book of Indian Birds* for the BNHS as its curator); Norman Boyd Kinnear (curator of the

With Loke Wan Tho: An abiding friendship

Society, who after the World War I joined the bird room of the British Museum (Natural History) and became Sir Kinnear and Director between 1947 and 1950); Colonel R. Meinertzhagen (noted British ornithologist who was a bit of a "bully" and with whom Salim Ali went on a birding expedition to Afghanistan); Hugh Whistler (who was in the Imperial Police Service and a keen observer of Indian birds and who gave "admirable prescriptions" on how to do a bird survey); and Sidney Dillion Ripley (Secretary, Smithsonian Institution, US, co-author with whom Salim Ali wrote his *magnum opus*, the ten volume *The Handbook of Birds of India* and *Pakistan,* and *A Pictorial Guide to the Birds of the Indian Subcontinent*). Ripley, a close friend and another giant of his age, was posted in Sri Lanka in the 1940s and he often visited the BNHS. His association with Salim Ali resulted in several joint ornithological expeditions and research collaborations.

Salim Ali tied the nuptial knot in December, 1918, to an "attractive relation" named Tehmina, the daughter of a pearl business man, C.A. Latif, who had migrated to London with his family, but had returned to Bombay in 1915. Though educated abroad and used to a sophisticated life, Tehmina "also loved the wooded countryside and jungle walks," and like her husband enjoyed good books and poetry, both in Urdu and English. And she happily participated in his camps, read his manuscripts (though she did not live to see his first book), and took a keen interest in birds. Until her unfortunate death from blood poisoning in 1939 after a small surgery in Dehra Dun, where the Alis had gone to live after returning from Burma, Tehmina remained most supportive of her husband's work, which initially involved many hardships and little money. "She encouraged me in every way in my work, which for many years was absolutely unremunerative," Salim Ali had stated. "She was so keen that I should do something that I had my heart in… If the wife had grumbled and said look at Mrs so and so, she has got a car and a frigidaire, it would have been impossible for me to work…" But in good time the Alis had their car named 'Jane' and a refrigerator too.

圖 運 陸

Loke Wan Tho

PHONE 3403/4

10 X/46

Cathay Cinema,

Singapore.

2nd October, 1946.

Salim Ali Esq.,
33 Pali Hill,
Bandra, Bombay.

My dear Salim,

I have today sent off on the American President Steamship the Louis McHenry Howe one sleeping bag and one tin trunk containing articles of clothing needed for the trip to the Mishmi Hills. The tin trunk contains the following articles:-

2 blankets
3 pieces khaki woollen trousers
2 " cotton trousers
1 " grey flannels
1 piece woollen khaki cloth
3 woollen singlets
2 " pants
3 " shirts
3 " pullovers
2 suits wool pyjamas
1 leather jacket
2 khaki wool shirts
4 pairs wool socks
2 pairs boots
1 black photographic changing bag
1 balaclava
2 pairs ankle puttees

From the ever helpful Loke Wan Tho

A dear friend, whom Salim Ali described as "a most unusual and lovable character" and with whom he remained very close until his tragic death in 1964 was Loke Wan Tho. A Singapore millionaire, cinema owner, a businessman and foremost a bird photographer, he was in exile in Bombay when the Japanese overran Singapore during the Second World War. Loke accompanied Salim Ali to numerous birding expeditions and international bird conferences and went with him to a number of places on bird photographing holidays. In a letter dated January 4, 1948, Loke asked as he often did, "How are you getting along for funds? Let me know when they run low as I may be able to put my hand to another sea green cheque." He also presented Salim Ali with money for buying a station wagon for his ornithological trips. In a letter dated March 29, 1948, Salim Ali wrote to his friend, "I cannot tell you what a blessing your station wagon proved all through the trip. It would have been quite impossible to move about the country as we did without it, and but for it I would never have reached many of the remote spots unless I had had three times as much time and 20 times as much energy to get to them by bullock cart. I think it was magnificent of you to have thought of the station wagon idea; I only wish you could make some use of it also. I feel an awful humbug driving about in style as if I owned it! I have bought a jeep trailer for the baggage and between them they make a most useful turnout."

A widely travelled man, Salim Ali's fieldwork took him from Kailash in the north to Travancore in the south and from Kutch in the west to Namdapha in the east, so that he was said to be like Jawarharlal Nehru, one of India's most travelled men. After returning from Germany and finding himself unemployed, he invented his own work by volunteering to do regional bird surveys on behalf of the BNHS without drawing a salary, asking the Society to organise funds only for camping expenses and to arrange for local help and provide a field collector and a taxidermist to assist him. Finally, these very surveys in the different parts of the country—Hyderabad, the Nilgiris, Bahawalpur, Bastar, Bhutan, Kutch, Sikkim,

Asad Rahmani

Travancore and Cochin made him the man he was: a walking encyclopaedia and the topmost authority on Indian birds. These surveys provided him with ample data for his bird books that brought him name, fame and also financial relief to pay off his old business debt.

In his later years, the BNHS, located at Dr. Salim Ali Chowk in Mumbai, a premier institute for the promotion of conservation and natural history, became alive with his magic. After independence, he was the Society's first Indian Honorary Secretary and the editor of the Society's Journal, and at the time of his death, its President. Due to his initiative, the University of Bombay came to recognise the BNHS as an institute for post-graduate research in Field Ornithology, leading to Master's and Doctoral degrees in zoology. He also initiated a fund under the BNHS called the Salim Ali—Loke Ornithological Research Fund (SALOR) with a seed donation, and the Salim Ali Nature Conservation Fund, which later received generous donations from his friend, Loke's family.

In his later years, Salim Ali, who saw the Halley's comet twice in 1910 and in 1987 and lived to be 91, developed a hearing problem. With generous donations from friends in the USA and England, and in spite of his protests against the 'fuss', a new set of hearing aids was obtained for him, which once again restored the sound of bird calls and birdsong to his ears. And though he grew frail with age, to the very end of his life, he lived a naturalist's tough life, showing no signs of slowing down.

Such was Salim Ali's commitment to his life's work that he was never ready to quit. "I am not ready to die," he once said. "I have only touched the surface." His last field trip was to Vansda National Park (Gujarat) in February 1987, when he travelled by car for 175 miles and he was well past 90. On June 20, 1987, Salim Ali passed away, having suffered for a period from prostrate cancer. The Birdman who, people said, looked like a bird—a white feathered bulbul? No, a vulture, he had said—has left behind a priceless legacy: a treasury of primary data on India's avifauna, which he painstakingly collected and which today forms the basic reference point of all new studies in Indian ornithology. Those of us who share with Salim Ali even a tiny fraction of his fascination with birds, cannot but remain indebted to the ornithologist extraordinary for his work with birds, the most fascinating and beautiful of our animal friends.

Chapter Two

Ornithologist Extraordinary

"My own interest in ornithology, as you may know, has always been in the living bird, its habitats, ecology and biogeography," so wrote Salim Ali to Professor Erwin Stresemann, his *guru*, in a letter dated September 27, 1970. "Throughout my ornithological career, I have devoted my special attention to these aspects, and I have derived considerable pleasure, profit and satisfaction from studying them."

For over seventy long years, the Grand Old Man of Indian ornithology observed birds under natural conditions and "not in a laboratory under a microscope," as he put it. By travelling to every corner of the country, to remote, uninhabited places, he studied birds "as they lived and behaved in their habitats." He climbed towers risking his life, almost fell off a precipice, rode on camel's back for hours in the salt desert of the Rann of Kutch, trekked for miles on mountaintops, and trudged through thick evergreen forests and scrubland just to be with the birds. Sometimes, travelling on foot or moving by bullock cart or by bus, he crossed the miles of dusty, bumpy, cross-country *kutcha* roads and makeshift bridges to see his birds. In his days, there were neither good roads nor the hardy vehicles we have today designed for wildnerness trails.

Once in Bastar, he undertook a "six hours non-stop marching," when the wheel of his vehicle came off and there was no means of sending a SOS in the middle of a thick forest, where a man-eating tiger had been reported. He described the experience as "eerie nerve-racking." Syed Aftab Hussain, who was close to the Old Man and who accompanied him in his travels in India and overseas has recalled, "In Ladakh, in search of the Black-necked Cranes, we had to walk on the marshes where you sink in if you do not keep moving quickly. We reported seeing the cranes to Salim Ali who was eager to go and see them. We put him on a horse but the wooden saddles they use there were too uncomfortable and so we decided to drive the cranes towards him. He put on a Ladakhi dress as camouflage..." All this, he enjoyed and was "fun" for him. In *The Fall of Sparrow*, his autobiography, he wrote

Asad Rahmani

about the joy of his work: "...I have watched birds through half a century and more, chiefly for the pleasure and elation of the spirit they have afforded. Bird-watching provided the excuse for removing myself to where every prospect pleases—up in the mountains or deep in the jungles—away from the noisy rough and tumble of the dubious civilisation of this mechanical high speed-age..."

The first Indian ornithologist to undertake several collection and observation surveys in order to document bird life of the Indian subcontinent, Salim Ali also wrote several bird books that became popular with the common man. And apart from these books that he wrote and the 36 or more radio talks that he gave on birds, he also lectured at several universities and clubs and wrote over 104 research papers on birds including topics such as *Courtship and Display among Birds, Bird Watching for Mountaineers, Bird Migration, Usefulness of Birds, Bird-Flowers and Flower-Birds in India—A Romance of Inter-linked Lives* (which he wrote for the *National Geographic* magazine in 1931) and so on. And wherever and whenever he had an opportunity, he advocated protection and conservation of our natural heritage.

After returning from Germany where he trained under Professor Erwin Stresemann at the Berlin University Zoological Museum for nine months, Salim Ali found himself searching in vain for his dream job in ornithology. Perhaps, with his scientific temper and his excellent command over the English language, he could have found a comfortable office job if he had so wished, but his interest lay purely in natural history. And as he did not have a formal degree, he found a job in his chosen line hard to come by. Yet, nothing could dampen his spirit and weaken his resolution to make birds the love and mission of his life, and so he decided to carry out private research in the vast, unexplored tracks of the country, particularly in the former Princely States of post-independent India, and record for posterity the avifauna of these regions.

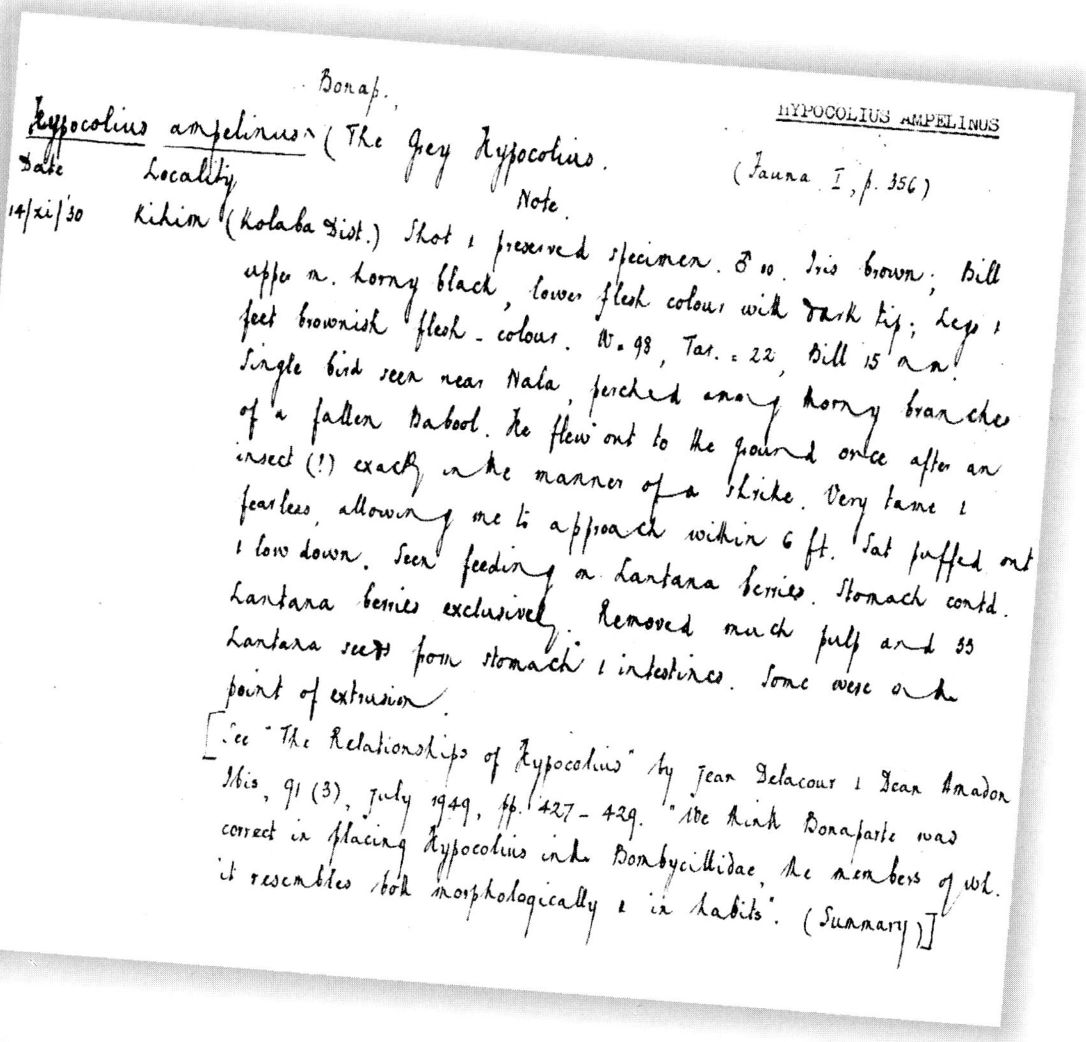

HYPOCOLIUS AMPELINUS

Hypocolius ampelinus Bonap. (The Grey Hypocolius. (Fauna I, p. 356)

Date Locality Note.

14/xi/30 Kihim (Kolaba Dist.) Shot & preserved specimen. ♂ 10. Iris brown; Bill upper m. horny black, lower flesh colour with dark tip; Legs & feet brownish flesh-colour. W. 98, Tar. = 22, Bill 15 mm. Single bird seen near Nala, perched among thorny branches of a fallen Babool. He flew out to the ground once after an insect (!) exactly in the manner of a Shrike. Very tame & fearless, allowing me to approach within 6 ft. Sat puffed out & low down. Seen feeding on Lantana berries. Stomach contd. Lantana berries exclusively. Removed much pulp and 33 Lantana seeds from stomach & intestines. Some were on the point of extrusion.

[See "The Relationships of Hypocolius" by Jean Delacour & Dean Amadon Ibis, 91 (3), July 1949, pp. 427-429. "We think Bonaparte was correct in placing Hypocolius in the Bombycillidae, the members of wh. it resembles both morphologically & in habits". (Summary)]

A page from Salim Ali's bird diary

Years later in an interview, Salim Ali recalled, "Finally, I said that well, I have these training and I have my chief interest in birds, so why should I not do this on my own. My wife had a little money and I had a little investment and so on. Then we worked out and found that we had just enough if we left Bombay, which was very expensive and went to live in some quieter place. Then I told Prater (then, the curator of BNHS) 'Look we have so many places in India and we know nothing about the birds.' Hyderabad for instance, was a complete blank on the ornithological map. So I said, 'If you write to the British Residents who are really interested in these kind of things, we can probably get some financial support. I do not want any pay. I only want my expenses paid and I will be quite happy to go, study and collect birds.' So, the Society got in touch with the Hyderabad Government, which had largely British heads of departments. They were very glad..."

Thus began Salim Ali's ornithological odyssey to the far-flung places of the country to study, collect, document and photograph birds. And with his collection and observation surveys covering the entire country from Kailash in the north to Kerala in the south, and Kutch in the west to the Mishmi Hills in the east and the literature he produced on birds, Salim Ali contributed greatly to ornithological knowledge in India.

Some of the important collection surveys (mostly sponsored by the Maharajas of yesteryears) that he undertook soon after returning from Germany in 1930 were:

- The Hyderbad State Ornithological Survey, 1931/2
- The Ornithological Survey of TravancoreCochin, 1933
- The Bahawalpur Survey of 1934/9 in the Rajputana States Agency
- Field Survey of Birds of Kutch, 1945
- Bird Survey of Gujarat, 1946

- First expedition to the Mishmi Hills, eastern Himalaya, 1948
- Bird Survey of Bastar, 1949
- Survey of Birds of Sikkim, 1955
- Ornithological Survey of Birds of Bhutan, 1966, 1973

Asad Rahmani

At a camp in Nannaj, Solapur, 1984

R.Dis.No.1113/32-E.

GOVERNMENT OF TRAVANCORE.
Huzur Outcherry.
Trivandrum, 3rd October 1932.

From

The Chief Secretary to Government,

To

The Honorary Secretary,
Bombay Natural History Society,
6, Apollo Street, Bombay.

Sir,

With reference to your letter dated the 18th August 1932, re: the Ornithological Survey by Mr.Salim A.Ali, I have the honour to inform you that the Government of His Highness the Maharaja have been pleased to sanction a grant of Bh.R[s] 2,000/-(two thousand) towards the cost of the Survey in Travancore.

2. The Superintendent, Government Museum and Public Gardens, Trivandrum, has been asked to cooperate in the work and, if possible to send one of his subordinates to join the Survey party on receipt of information from the party.

3. Instructions have also been issued for the Traveller's Bungalows, Campsheds, and Forest Rest Houses being placed at the disposal of the Survey party free of rent and for Government tents being given to the party on loan.

4. The Commissioner of Police, the Land Revenue and Income Tax Commissioner and the Conservator of Forests have been asked to arrange for the necessary assistance being rendered to the party on their requisition.

5. Government note that the State will have a claim on the collection made as a result of the Survey and that a complete set of specimens will be presented to the State.

I have the honour to be,
Sir,
Your most obedient servant,
Sd/ T.R.Muthuswamy Aiyar
for Chief Secretary to Government.

The Maharaja's sanction letter for the Tranvancore Survey

The Hyderabad survey was the first collection survey. After this, Salim Ali recalled, "I did Kerala, which was then two states, Cochin and Travancore. Then one after the other: Central India, Gwalior, Indore, Bhopal."

The Travancore-Cochin survey report was highly commended in scientific journals and it gave the ornithologist the "greatest satisfaction". Hamid Ali, Salim Ali's brother, who had encouraged him to develop as a naturalist, described the five-months Travancore-Cochin survey carried out in the midst of the evergreen forests of Kerala with its extraordinary richness and diversity of bird life as "the most thorough in content and matter."

Another significant survey was the Bahawalapur survey. Prior to this study, hardly any data existed on the bird life of the Rajputana desert. This survey resulted in a fascinating study of "camouflage colouration of desert animals." It had been observed earlier that low humidity in desert atmosphere, which resulted in high ultraviolet radiation, caused the paleness of desert soil and animals. Salim Ali was the first to observe the reverse phenomenon: that an increase in humidity after a shower, which lowered radiation resulted in darkening the colour of desert birds, like that of the Desert Finch-Lark (*Ammomanes desert*) and the Desert Courser (*Cursorius cursor*), "to exactly the same shade of brown as the sodden soil upon which they found themselves."

The Kutch survey was undertaken at the invitation of Maharao Vijayarajji of Kutch. Though an extremely difficult terrain, Salim Ali described the Rann of Kutch, the largest breeding colony of the Greater Flamingo (*Phoenicopterus roseus*) in Asia and one of the largest in the world, as "my favourite stamping ground because the Flamingo City is there..." Here, for the first time, he made "a fairly accurate physical count" (about half a million) of these "long-legged, long-necked rosy white stork-like birds" during their peak breeding season. He also observed their nocturnal movements and behaviour and the geographical conditions—a certain depth of water on the breeding grounds necessary for the birds to

build their nests. Here Salim Ali recorded the presence of a colony of Avocets (*Recurvirostra avosetta*) outside the Flamingo colony and a nesting colony of Rosy pelicans among the disused Flamingo nests.

An important discovery of the Bastar survey was the sighting of the Great Black Woodpecker (*Dryocopus javensis*), locally known as *Bhainsa-khidree*, in central India. "The extension of the range of the Great Black Woodpecker, and some other hitherto considered predominantly Western Ghat forms, right across continental India to the Eastern Ghats, is among the significant findings of the Bastar/Kanker bird survey," he noted in his memoir.

After the Bastar survey, several other field studies were undertaken covering vast tracts in Madhya Pradesh, Andhra Pradesh and Orissa. He also undertook joint ornithological expeditions with S.D. Ripley, the noted American ornithologist, to Simlipal, Orissa, Melghat, Maharashtra, Arunachal Pradesh and Bhutan, which provided extensive field data for the ten volume *The Handbook of Birds of India and Pakistan*, which they co-authored.

The most significant and path breaking bird studies for which Salim Ali is known was on the extraordinary breeding biology of the Baya Weaver birds (*Ploceus Phillippinus*). "The best, the longest and most unbroken opportunity I had (to study) was on the weaver birds," Salim Ali said. The paper published in the Journal of the BNHS, was one of the firsts in the world on the behaviour of wild birds. Until then, research was centred on morphology, and behavioral research based on watching birds for hours together in the field was not known.

Klaesson

With The Handbook of Birds, 1983

H.R. Yadav

The extraordinary Bayas

On returning from Germany in 1930 and finding himself jobless, Salim Ali had retreated with his wife, Tehmina, to her family house in Kihim outside Bombay, where living would be less expensive and he could carry on with his bird-watching.

Here, during the monsoon, he found a nesting colony of Baya Weaver birds. The bayas live in colonies and their bottle-shaped nest that dangles from the branches of babul and palmyra trees is a common site in the countryside. But little was so far known of their breeding behaviour till Salim Ali observed that only the male baya builds a nest and that he acquired two to five 'wives'. When the nest is half built, the female goes from nest to nest inspecting them, while the male waits for her to make her choice. Once, a female accepts a nest, the two pair and eggs are laid. But if the nest is rejected, the male promptly builds another. And once a female has laid her eggs, the polygamous male moves on to build another nest for a new 'wife' and the process is repeated. The baya study established once for all the importance of field study in ornithology.

Another important project that was close to Salim Ali's heart was the study of the fascinating phenomenon of bird migration. With its headquarters at Bharatpur, the project was launched with the help of the World Health Organization, which was keen to find out the possibility of virus travelling through bird ticks by way of migration. The study involved mass ringing of birds and for this camps were set up in all the important locations, including the Great Rann of Kutch, which receives a large number of migratory birds coming from Europe, Siberia and north and central Asia.

Richard Clark

At the family house in Kihim

In the 1930s and the '40s, ornithological surveys were based on collection of specimens and on visual observations. Technology like satellite tracking, telemetry, etc. did not exist. "And yet," says Prakash Rao, WWF, India, "it is remarkable how Dr. Ali got the same results and achieved so much without the aid of technology that modern researchers today have access to. For him, getting information out of the local villagers was important. He insisted with his students on winning the confidence of the villagers."

All his life Salim Ali looked for birds, studied them and searched for the lost ones. He had looked for the Mountain Quail (*Ophrysia superciliosa*) of the Kumaon Himalaya last reported in 1876; the Blewitt's Owl or Forest Spotted Owlet (*Athene blewittii*) of the Satpura highlands last spotted in 1914 in Orissa; the Jerdon's Courser

Mountain Quail

H.R. Yadav

(*Cursorius bitorquatus*) of the Deccan plateau last seen in 1900; and the Pinkheaded duck *(Rhodonessa caryophyllacea)* of Bengal/Bihar, which he saw not in the wild but 11 of them in the duck pond of one Alfred Ezra in London. He had a photograph of these ducks, which he said was "a very bad photograph, but perhaps the only one of the living Pinkheaded Duck."

Describing his search for the Mountain Quail, Salim Ali said, "I have not seen the Mountain Quail although we have beaten up and down the same hills where they were collected dozens of times. People still claim they were there and they have seen them." There are only six specimens of these birds in the world collected in 1879 (four in Mussoorie and two in Naini Tal), after which they vanished. Salim Ali tried to locate these lost birds with the help of dogs by flushing them out from bushes, but the searches yielded no result. He also looked for the rare Lesser Florican (*Sypheotides Indica*), locally known as *Kharmor*, several times in Indore, which he had first seen in 1930

Jerdon's Couser

Pinkheaded Duck

when it was abundant till the 1960s, and which he again looked for in 1981 and 1984, and later a study was started on this bird under his guidance.

Today, the search for the Mountain Quail is still on; the forest spotted owlet was finally found in 1997, ten years after Salim Ali's death; and the Jerdon's Courser was rediscovered in 1986 in Andhra Pradesh after several attempts to find it had failed. A reason given for earlier failure in finding the bird is that the search was conducted wrongly during daytime. The courser is nocturnal and comes out to feed in the evening, a time when many locals had seen it.

Salim Ali and his team has been credited with the rediscovery of the Honeyguide, which was not heard of or seen for at least 150 years but was found during the Bhutan expedition. This bird belongs to a family that has about 13 species in Africa and one in the Himalaya and another in Vietnam. It is said to feed on beeswax, not known to be eaten by any other bird. Salim Ali also discovered several new subspecies of Indian birds, i.e. different ecological races that look alike except for slight physical and behavioral variations.

White Eye

And some of the birds discovered by other ornithologists were named after Salim Ali. These were the Oriental White-Eye (*Zosterops palpebrosa salimalii, 1933*), a bird from Hyderabad which sucks nectar from flowers like a hummingbird; the Rock Bush Quail (*Perdicula argoondah slimalii, 1943*), a Brick-Red Quail that lives on brick-red soil in Mysore; the Hodgson's Scimitar Babbler (*Pomatorhinus horsfieldii salimalii, 1948*) from Tezu, Mishmi Hills; the Finn's Weaver (*Ploceus megarhynchus salimalii, 1959*) from Bhutan; the Pacific Swift (*Apus pacificus salimalii, 1958*) from southeast Tibet; the Goldcrest (*Regulus regulus salmalii, 1954*) from Simla Hills; the Emerald Dove (*Chalcophaps indica salimalii, 1960*) from Tenmalai, Kerala, and the Streaked Fantail Warbler (*Chalcophaps indica salimalii,*

1936) from Peerumedu, Kerala. An ecological race found in Kerala, which is related to the Golden-backed Woodpecker (*Dinopium benghalense*) and is common in many parts of India was named *Dinopium benghalense tehminae* after his wife, Tehmina.

Golden-backed Woodpecker

Salim Ali's eminently readable books on Indian birds, as noted ealier, have helped to make the average Indian become aware of birds in one's surroundings. "I firmly believe that besides providing factual scientific information to the reader it is just as important to make that account pleasurable reading..." noted Salim Ali. *The Book of Indian Birds,* which has sold thousands of copies and has been translated into several regional languages, was read by Jawarharlal Nehru, Maulana Abul Kalam Azad, and Indira Gandhi during their prison days. In his Azad Memorial Lecture (1978), *Bird Study in Indiaits History and its Importance*, Salim Ali recalled:

"I had been reading in camp with much delectation and many a chuckle, Maulana Azad's inimitable essay *Chirya Chirya ki Kahani* in his admirable Urdu classic *Ghubar-e-Khatir* written during his long incarceration in Ahmadnagar jail.... In describing the nest-feeding activities of the parent birds in his cell, the Maulana quotes the findings of "an ornithologist" on the average number of daily visits a pair of sparrows makes with food for the nestling. When I first read the account, I liked to identify that ornithologist as myself, and to fancy that the information may have been culled from my own *Book of Indian Birds*. Talking to his personal secretary... years later, I learnt that my surmise was indeed correct. He had got his statement out of a copy of the above book lent to him in Ahmadnagar Fort by his fellow prisoner Jawaharlal Nehru—himself a keen bird student and nature lover!"

Eminently readable books on Indian birds

The Book of Indian Birds (1941) was followed by *The Bird of Kutch*, (1945) and the *Indian Hill Birds* (1949) all based on his surveys and field studies. The Travancore-Cochin survey too resulted in a book: the *Birds of Kerala* (third reprint 1985). Next followed *The Birds of Sikkim* (1962), and a similar book was planned for Bhutan when he visited the Himayalan State for a bird survey at the invitation of the King of Bhutan, Druk Gyalpo Jigme Dorji. But as a great deal of similarity exists in the bird life of Bhutan, Sikkim and Arunachal Pradesh, a book covering the entire region*The Field Guide to the Birds of the Eastern Himalayas* (1977) was published. The first volume of the ten volume *The Handbook of the Birds of India and Pakistan* by Salim Ali and S.D. Ripley was published in 1968 and the last in 1974. The series took some thirty long years of work and a compact edition incorporating all the ten volumes appeared in 1983. The authors had thought of this work during their birding trip to the Mishmi Hills of northeast Assam and for this project Salim Ali put in 12 to 15 hours of work a day, writing the first draft by hand.

Ripley, who passed away in 2001, said in a tribute to his friend, "It is rare to find a long-time friend with whom one can wholeheartedly pursue shared goals that result in advances in both science and conservation. Such a relationship, based largely on mutual respect and shared interests, existed between Dr. Salim Ali and myself, and lasted from 1943 through Dr. Ali's passing in 1987."

An ardent naturalist, in a talk addressed to mountaineers, Salim Ali appealed to them to use the unique opportunity they had to join hands in the study of natural history. He said, "I come to you with the old grouse, that is to say that I am very disappointed in mountaineers who just climb a mountain and come down and bring us no information on natural history, flora and fauna they see. So, I feel, that those of you, who are at all interested, may write to the Bombay Natural History Society before starting on an expedition. We would be very happy to put you at least on the way of observing intelligently..."

So complete was the Birdman's interest in birds that he studied anything and everything to do with birds. He took a keen interest in Moghul miniatures, which have excellent representations of Indian birds. His three part study *The Mughal Emperors of India as Naturalists and Sportsmen*, (published in the Journal of the BNHS, 1927/28), has been described as a pioneering work of an ornithologist in the field of art and history. Interestingly, referring to the Jahangir School of animal paintings in his Azad Memorial Lecture, he observed: "Among the bird portraits attributed to Ustad Mansur... is a painting of a Siberian or Great White Crane. At the present time this crane is one of our rarest and most endangered species... (which) visits the Keoladeo Ghana Bird Sanctuary of Bharatpur every winter between November and March to add to the glory of this priceless natural wetland. The interesting point about this painting—fortunately still in the collection of the Indian Museum, Kolkata, is that it was executed between the years AD 1616 and 1620 and thus, nearly 160 years before the bird was actually 'discovered' and described for western science (in 1773) by the Russian zoologist Pallas from the swamps bordering the Ob and Irtysh rivers in the USSR!"

Siberian or Great white Crane

The scientist-naturalist, Salim Ali was also one of India's first and foremost conservationists. He was made the Honorary member of the International Union for Conservation of Nature and Natural Resource in 1981 for his outstanding contribution to nature conservation. After the Second World War, when the conservation movement took root and the Indian Board for Wildlife (IBWL) and various sanctuaries and national parks came to be formed, he and the BNHS played a major role in the process. As S.D. Ripley explained, "At the start of our acquaintance, conservation was not our major concern; however, after revisiting our field sites and those of others, we could not fail to note with sadness the inexorable habitat destruction and the accompanying loss of biodiversity in once-pristine areas... Later in our careers, we became increasingly motivated by the overriding importance of conservation in India, and we expended considerable time and energy speaking out on the critical need for conservation."

As a matter of fact, the Keoladeo National Park at Bharatpur, a unique wetland and a major centre of ornithological research, owes its existence to Salim Ali's efforts in saving it from being vandalised. In this, his friendship with Jawaharlal Nehru since his Dehra Dun days, when he visited him in jail, paid dividend. In a letter to Indira Gandhi, dated March 10, 1969, he wrote: "I had occasion to resort to similar pleading with your father when, in the early flush of independence, local politicians having scores to settle with the Maharaja of Bharatpur, seemed bent on 'punishing' him by destroying the Keoladeo Ghana Jheel.... It was only with Panditji's personal interest and ready support that we (Horace Alexander, General Williams and myself) were then able to save the jheel from imminent doom and to get it notified as a sanctuary for breeding waterbirds by the Rajasthan Government. ...We would like to be... assured that under no circumstances will anything be done or allowed to be done to damage or destroy this fabulous waterbird resort, the like of which it would be difficult to find anywhere in the world... To any visitor with an eye for Nature, the Keoladeo Ghana Bird Sanctuary is certainly as great an attraction as the Taj Mahal..."

Salim Ali had also lent his support to the Silent Valley campaign in Kerala and was instrumental in influencing Indira Gandhi to shelve the hydro-electric project that would

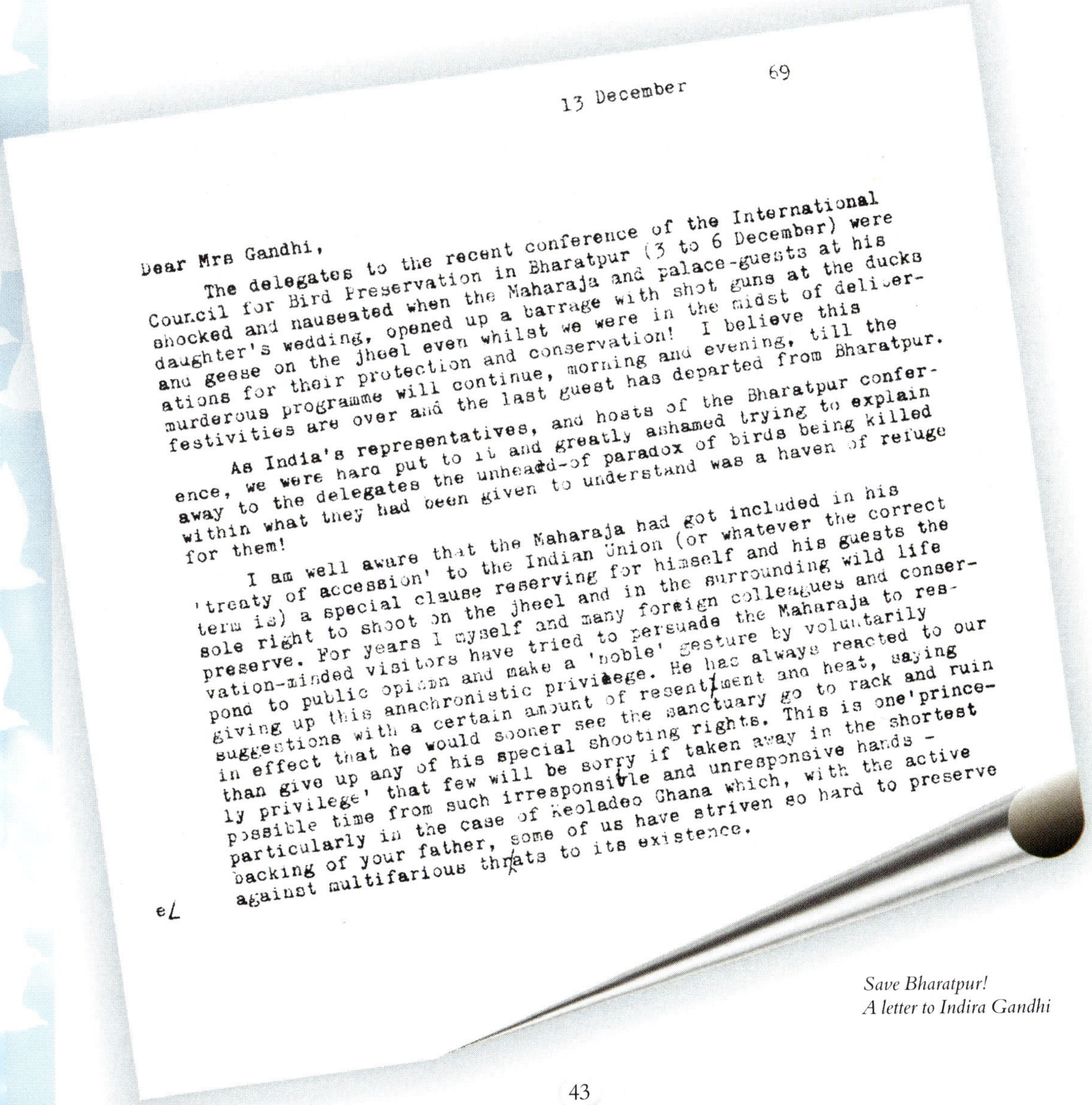

13 December 69

Dear Mrs Gandhi,

The delegates to the recent conference of the International Council for Bird Preservation in Bharatpur (3 to 6 December) were shocked and nauseated when the Maharaja and palace-guests at his daughter's wedding, opened up a barrage with shot guns at the ducks and geese on the jheel even whilst we were in the midst of deliberations for their protection and conservation! I believe this murderous programme will continue, morning and evening, till the festivities are over and the last guest has departed from Bharatpur.

As India's representatives, and hosts of the Bharatpur conference, we were hard put to it and greatly ashamed trying to explain away to the delegates the unheard-of paradox of birds being killed within what they had been given to understand was a haven of refuge for them!

I am well aware that the Maharaja had got included in his 'treaty of accession' to the Indian Union (or whatever the correct term is) a special clause reserving for himself and his guests the sole right to shoot on the jheel and in the surrounding wild life preserve. For years I myself and many foreign colleagues and conservation-minded visitors have tried to persuade the Maharaja to respond to public opinion and make a 'noble' gesture by voluntarily giving up this anachronistic privilege. He has always reacted to our suggestions with a certain amount of resentment and heat, saying in effect that he would sooner see the sanctuary go to rack and ruin than give up any of his special shooting rights. This is one 'princely privilege' that few will be sorry if taken away in the shortest possible time from such irresponsible and unresponsive hands - particularly in the case of Keoladeo Ghana which, with the active backing of your father, some of us have striven so hard to preserve against multifarious threats to its existence.

Save Bharatpur!
A letter to Indira Gandhi

Migratory birds on flight to Keoladeo Ghana Bird Sanctuary: As great an attraction as the Taj Mahal

have drowned hundreds of acres of prime forest land. Mrs. Gandhi announcing the IBWL's first gold medal for conservation awarded to Salim Ali honoured the Grand Old Man with a befitting tribute: "For 50 years Salim Ali has been the doyen of conservation in the country. He is responsible for the new consciousness sweeping the country. He may be the oldest member in the Society but there are few younger in heart or as energetic."

Chapter Three

The Birdman as a Man

In 1996, the Bombay Natural History Society (BNHS), with which Salim Ali was so closely associated, celebrated his birth centenary by dedicating to him a special issue of their *Hornbill* magazine to felicitate the Grand Old Man. A pen-portrait of 'Doctaar Sahib', as Salim Ali came to be called in the far-flung corners of country from that issue:

"A lifetime dedicated to the study of birds, a fine teacher, an exacting critic and beloved friend, a mentor for several generations of nature lovers and biologists and father figure for BNHS, the institution that he nurtured and to which he gave of himself completely."

Describing the ornithologist's passion for birds (a passion sans sentimentality), J.C. Daniel, the present Honorary Secretary of the BNHS, who had the longest number of years of association with Salim Ali, wrote in one of its publications *Salim Ali's India*: "In those days, anybody with his energy and command of English could easily have found a comfortable niche in a Bombay firm, but Salim Ali didn't even bother to apply. His family thought him mad, an unemployed married man, cheerfully content to watch birds all the time." But even in his lifetime, Salim Ali became something of a living legend and fame and recognition were all his after the early years of struggle and hardship were over. His pioneering study of birds of the Indian subcontinent earned him the highest esteem of his colleagues and associates across the world.

Today, if amateur bird-watchers can differentiate between a Bank Myna from a Common Myna, or a Cattle Egret from a Little Egret and squeal with delight on recognising other birds and rattle off their names—barbet, chloropsis, bushchat, shoveller, sheldrake, wryneck, etc.—it is because of Salim Ali's accurately illustrated reader-friendly bird books, which have helped to popularise bird-watching as an enjoyable pastime among the people. "Bird-watching is not only interesting, but also an enjoyable hobby," he said. "All that you need is an inexpensive pair of binoculars, a notebook, a pencil and an ample stock of patience and dedication."

And patience and dedication the Birdman had in plenty. Rising early like the birds and sleeping lightly like any naturalist, he could put in 16 to 18 hours of long, grueling fieldwork, showing no signs of tiring. And even after the day's work was done, he would read aloud poetry (he was fond of English and Urdu poetry) from an anthology, often in the company of his friend, Loke Wan Tho, who shared with him a love for literature. Or, when in camp with fellow ornithologist, Sidney Dillion Ripley, with whom he shared an abiding friendship, the two would regale others with their exciting, humorous conversation sitting by a petromax lamp.

If the love of his life was birds, "humanity was his religion," says Asad Rahmani, present Director of the BNHS. "Religion was not important for the Old Man. For him to become a good human being was most important and the most difficult thing." Rahmani, who has studied the great Indian Bustard and the lesser Florican, both gravely endangered today, says, "I've been in the field with him many times (Sholapur and Sailana) and all the time he would be asking questions—the names of trees and flowers and butterflies. To use the simile 'curious as a child' is cliché, but he was really so. I used to correspond with him since I was a boy. And however busy he was, he would always reply to letters within ten days, whether it was from the Prime Minister or a person from a remote village or a school kid, the latter being most important for him."

Salim Ali's many students remember him as an incomparable teacher, even though he is said to have been exacting and with an irascible temper, of which he was well aware. J.C. Daniel, reminisced a year after his death: "Seeing him in one of his moods of frustration at failing to get his staff to achieve, reminded one of the temper tantrums of

Pandit Nehru at the shoddiness that surrounded him..." He also described him as a "tough and determined perfectionist," who could not bear "the general national attitude of *jaane do*..."

Indeed, Salim Ali had a great commitment to excellence; his hold on the English language was extremely good; his handwriting fine (he was upset if his students didn't have good handwriting and he advocated 'handwriting tests'); and "his equipment, though ancient, was spotless." He took down endless meticulous notes on the birds he observed throughout his life (his earliest bird note was written when he was a boy) and his over a dozen bird diaries make fascinating reading. Says Rahmani, "He had the correct word for every bird behaviour. You cannot find a more appropriate word than the one he used..." Adds Rajat Bhargava, Rahmani's student, "Dr. Salim Ali's observations and descriptions of bird behaviour and their habitats were minute, accurate and so detailed that even after half a century that they were written, he remains the foremost guide in the field." The young ornithologist, who is studying Finn's Baya was a bird keeper but he decided to take to field study of birds after reading J.C. Daniel's fine write-up *Unforgettable Salim Ali* that was written after the Grand Old Man passed away.

To the end of his life, Salim Ali continued to guide the young in ornithological research, while he himself carried on with his fieldwork, even when he was well past 90. He was always glad to meet young men or women with a love for the wild and the wilderness. Once, he said to a young wildlife photographer, "I am now too old and the field vast; one man's life is too short to complete the work. Therefore, I want that youngsters like you should give a helping hand, so that the work which I could not finish in my lifetime is carried on." Following the news of his demise, Rajiv Gandhi, then Prime Minister of India, said, in a tribute to the ornithologist, "He has died with the satisfaction of knowing that a large number of professional and amateur naturalists share his passion for preserving the wonderful variety of nature's manifestation."

A man of few needs, Salim Ali was frugal and spartan in his lifestyle and very careful in handling grants and funds. It is said, "he accounted for every rupee spent," and was very conscious of spending it rightly, returning even part of the grant than indulging in what he called "unnecessary comfort." Personally, he cared little for wealth and riches, and though later in life he earned a tidy sum from the sale of his books and the J. Paul Getty International Prize for Wildlife Conservation worth $50,000, he happily gave away a lion's share of the prize money to the BNHS to set up the Salim Ali Nature Conservation Fund for conservation and research.

Always rational, logical, meticulous and systematic in whatever he did, Salim Ali did not hesitate to question his own observations. Though gifted with sharp senses—a quick eye and remarkable ears for bird calls—he was always cautious when identifying a bird, and refrained from coming to hasty conclusions. This, he expected of his students as well, whom he did not hesitate to confound with an unnerving 'Are you sure?' query thereby, teaching them to "give every bird a second glance" and to scrutinise, rethink and re-examine facts. "Are you sure, it is the Jerdon's Courser?" he had asked Bharat Bhushan, the man credited with the rediscovery of the bird in Cuddapah in Andhra Pradesh in 1986 that had made international ornithological news. "If you are wrong, I will not bother to hang you from the nearest palm tree. You should do it yourself!" he had stated before expressing his delight and appreciation.

If his students honoured him and longed to be noticed by him, they also feared his "tongue-lashing" if ever a slight injury was caused to a bird caught in a mist net (first used in bird surveys in 1959, when Japanese mist nets were first introduced, which greatly helped in surveys and collections), or if the costly net got damaged. One of his students recalled, "He was a tough task master in camp and followed a strict routine. We, his students would make mumerous mistakes and get jittery. But he would never accept anything

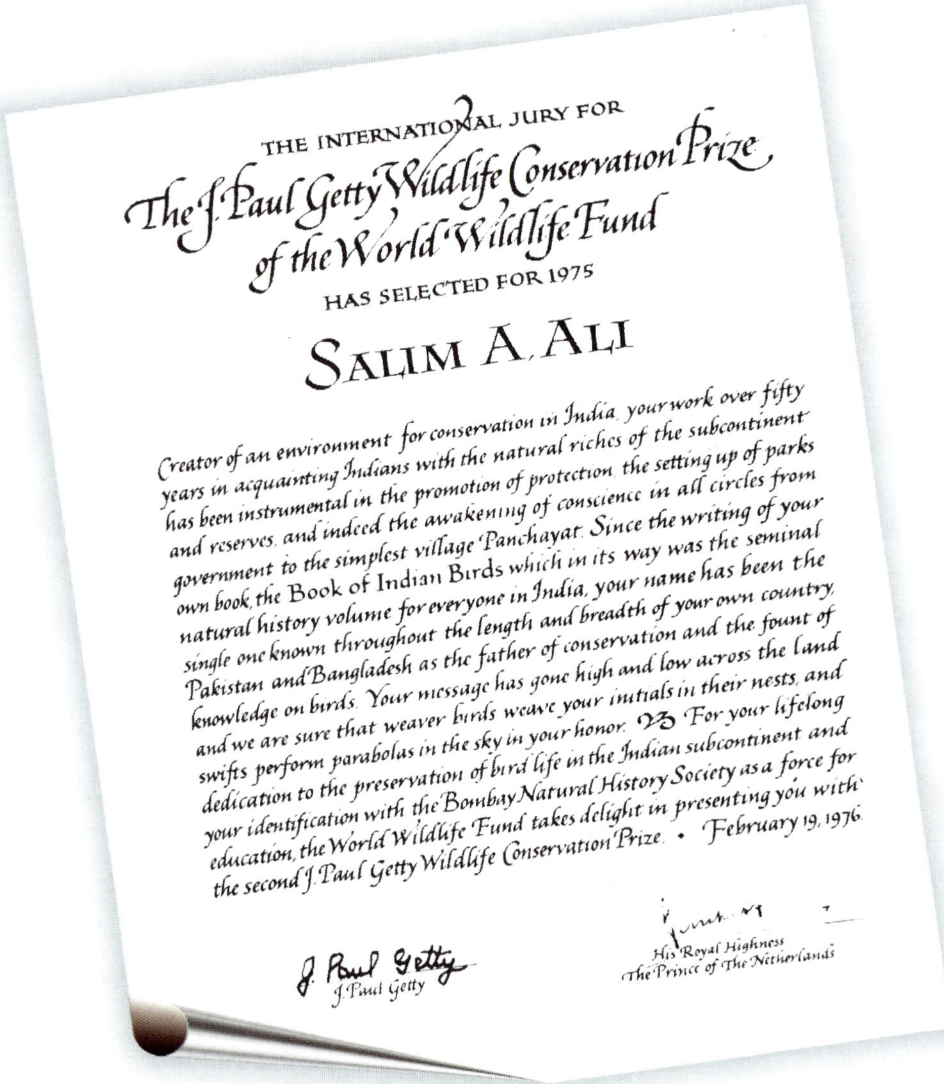

THE INTERNATIONAL JURY FOR
The J. Paul Getty Wildlife Conservation Prize
of the World Wildlife Fund
HAS SELECTED FOR 1975

SALIM A. ALI

Creator of an environment for conservation in India, your work over fifty years in acquainting Indians with the natural riches of the subcontinent has been instrumental in the promotion of protection, the setting up of parks and reserves, and indeed the awakening of conscience in all circles from government to the simplest village 'Panchayat. Since the writing of your own book, the Book of Indian Birds which in its way was the seminal natural history volume for everyone in India, your name has been the single one known throughout the length and breadth of your own country, Pakistan and Bangladesh as the father of conservation and the fount of knowledge on birds. Your message has gone high and low across the land and we are sure that weaver birds weave your initials in their nests, and swifts perform parabolas in the sky in your honor. For your lifelong dedication to the preservation of bird life in the Indian subcontinent and your identification with the Bombay Natural History Society as a force for education, the World Wildlife Fund takes delight in presenting you with the second J. Paul Getty Wildlife Conservation Prize. • February 19, 1976.

J. Paul Getty
J. Paul Getty

His Royal Highness
The Prince of The Netherlands

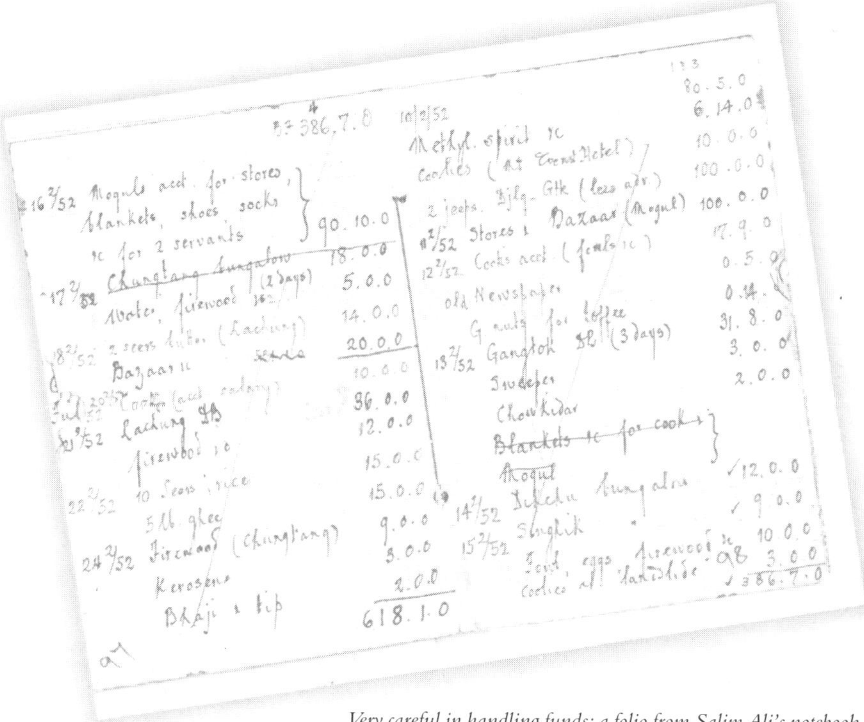

Very careful in handling funds: a folio from Salim Ali's notebook

sub-standard. If we prepared a specimen Salim Ali's way, it became perfect. And he made sure we did it. He was hundred per cent for quality."

In a letter to Erwin Stresemann, dated 16 April 1970, Salim Ali wrote candidly, "Queries and criticisms from the *guru* are always welcome and greatly appreciated. Only then, is it possible to try and rectify errors and clarify ambiguities." Among his students who did their M.Sc. and Doctorate under his guidance were V.C. Ambedkar (Bayas); V.S. Vijayan, current Director of Salim Ali Centre of Ornithology and Natural History, Coimbatore (Bulbuls); Robert Grubh, current Director of IRNE, Nagercoil (Vultures);

D.N. Mathew, Calicut (ecology and economic status of *Ploceus Philippinus* in agricultural environment); Priya Davidar (pollination of birds); H.S. Yahya (Barbets); Tara Gandhi (birds of the eucalyptus plantations); and Ali Raza Khan (Black and Orange Flycatcher) and Anwar-ul-Islam (Laughing Thrush), both from Bangladesh.

A strict disciplinarian, he blew his top when his students failed to arrive on time or fell behind schedule. But he could also be extremely caring and thoughtful towards the members of his expedition. "So much so that we had a policy not to show that we were sick because it invited all his attention and we did not want that," says S.A. Hussain, who joined the BNHS in 1969 and had worked with the Old Man in the field as well as accompanied him to his camps in Melghat, Simplipal, Bhutan, Andamans, Bangladesh, Sri Lanka, Point

Shahid Ali

House at Pali Hill, Mumbai

Hornbill

Calimere, Rann of Kutch, Harike Lake, Bharatpur, and Ladakh. Salim Ali was also a man of great wit and humour. Says S.M. Nair, former Director, National Museum of Natural History (NMNH), New Delhi, "He often used to visit the NMNH in the 1970s and the '80s in its formative years. Once he asked me where I got the specimens of mammals for the museum. I told him that the skins were acquired from the zoo whenever an animal died. Salim Ali laughed and said, 'That's good! So, if you want to kill an animal, send it to the zoo!"

Recalling the lighter side of the Birdman, Nair narrates another anecdote, which the old man had described with great amusement. Salim Ali was to visit Bhuj in Gujarat and so the BNHS sent a telegram to the Collector, Bhuj, stating that Salim Ali and a field collector from the BNHS would be reaching by such and such flight, so please receive him. The Bhuj office however, mistook the message as 'Collector from Bombay'. So, they made elaborate bandobost with VIP security arrangement... But to their consternation, the mistake was detected only after the Birdman had alighted from the plane in his khaki shorts, carrying his binoculars!

Fond of riding his motorbike and driving his vehicle, Salim Ali would drive down from his house at 46 Pali Hills in Bandra to Hornbill House, the BNHS office, even at the age of 78. "I started riding a motorcycle in 1915. Ever since I've kept it one after the other. I've had six or eight different makes, different models," he said and he regretted not possessing a BMW.

Childlike in his love for *barfis*, chocolates, *chikki* (peanuts in jaggery) and ice-creams, he once asked for a second helping of ice-cream just when the dessert plates were being cleared off the table. He was also known to have been fond of fruits—mango and

Perseverance and persistence to the end of his life

cherimoyer, a variety of custard apple, a native of Peru but now a Nilgiri specialty. And, like a child, he loved the outdoors, even preferring to sleep out at night on the verandah whenever he could. His colleagues and associates marvelled at his extraordinary energy and his sustained physical stamina, even in his late life, when he continued to take part in ringing birds and putting up nets.

Recalls Nair, "Whenever he used to come to the FICCI building that houses the NMNH, I noticed, he used to climb two steps at a time!" and if he ever hurt himself, as he once injured his back in the field, his love for nature took precedence over pain and discomfort. His friend Ripley, recalling the Old Man's rare spirit and power, said, "Our last expedition with Salim Ali was to Arunachal Pradesh, when he was about 90 years old! Even through the mid-1980s, we were still making plans for joint fieldwork; this resulted finally in a major ornithological survey of Namdapha in 1988, though Salim did not live to see it."

Once, a journalist wanted to know if the Birdman fancied acquiring wings so that he could fly like a bird. Amused by the naïve query, the naturalist replied, "My attitude towards the winged animals is not romantic." Adds S.A. Hussain, "The Old Man really hated mushy sentimentality. A bird was appreciated whether sitting on a tree for its sheer life, or on a dinner plate for its delicacy or as a museum specimen for scientific study." An 'ecologist with gun' as he was sometimes called, Salim Ali had to sacrifice "the lives of thousands birds," for his specimen collections, sometimes with "a prick of conscience," as he said. However, it was this painstaking and methodical collection that helped to build scientific knowledge of Indian birds as well as record their geographical distributions. "For a scientific approach to bird study, it is necessary to sacrifice a few," he said of his time. But as noted earlier, the focus of his study, as of all research today, was the living bird and bird behaviour in their ecological surroundings.

It is said, behind every successful man there is a woman. Salim Ali would have little quarrel with that, for he was always grateful to his wife, Tehmina, for her constant support in his career, until her untimely death. A widower for over four decades, he did not remarry and only once briefly fell in love. Hugh Whistler, the authority on Indian birds, writing to Salim Ali on 3 August 1939 at the news of his wife's death, described the couple as 'very, very devoted' and Tehmina as 'a rare companion' with which few men are blessed. At an interview, Salim Ali acknowledged, "My great fortune was my wife, who had had all her education in England and (had) been used to quite a different sort of life to what she would have in the kind of work I wished to do. She insisted that I should take up only the work that I was interest to do… Now when I look back, I think the chief of the 'chiefest' factors that made me continue with ornithology was my wife because you really cannot do much if you do not have a like minded companion."

With Salim Ali's death in 1987, India lost a great naturalist and a man who was refined, cultured and a patriot. "A man who was above all of us," says Rahmani, himself a renowned ornithologist. "He had no tolerance for sloppiness and disliked discussions, unending meetings, joint meetings, sub-meetings, which thoroughly bored him, when he is said to have quietly switched off his hearing aids! 'Go and do your work in the field,' he would say. And he believed if something could be said in two minutes, never waste two hours. Time was the most precious thing in life."

Honesty, perseverance, persistence, and hard work—these were the rare qualities that made Salim Ali the man he was, a man whose golden footsteps in the field of Indian ornithology shall forever shine.

New Names of Birds

Yellowthroated Sparrow—Chestnut-shouldered Petronica

Mountain Quail—Himalayan Quail

Great Black Woodpecker—White-bellied Woodpecker

Oriental White-Eye—White-Eye

Yellownaped Yuhina (*fuhina flavicollis*)—Whiskered Yuhina

Jerdon's Courser (*Cursorius bitorquatus*)—Jerdon's Courser (*Rhinoptilus bitorquatus*)

Great Indian Bustard—Indian Bustard

Golden-backed Woodpecker—Common Flameback

Pacific Swift—Fork-tailed Swift

Tibetan Crane—Black-necked Crane

Desert Courser—Cream-coloured Courser

References

1. *The Fall of a Sparrow,* Salim Ali, Oxford University Press,1985

2. *The Book of Indian Birds,* Salim Ali, BNHS/OUP, 1996

3. *A Pictorial Guide to the Birds of the Indian Subcontinent,* Salim Ali & S. Dillon Ripley (with 106 plates depicting all the birds by John Henry Dick), BNHS, Centenary publication

4. *Memories of Salim Ali,* Hornbill Magazine, 1988 3/4, BNHS

5. *Bird Study in India. Its History and its Importance,* Azad Memorial Lecture/delivered by Salim Ali, organised by the Indian Council for Cultural Relations

6. *Dr. Salim Ali Papers,* Nehru Memorial Library, New Delhi

Acknowledgements

*My grateful thanks to Dr. Asad R. Rahmani, Director,
Bombay Natural History Society, for permission to use
photographs from Hornbill and for his valuable time in going through
the manuscript. To Rajat Bhargava for all his help.
To S. A. Hussain and Dr. Prakash Rao, WWF, India
for reading a chapter each.*

NOTES